INDIAN VEGETARIAN Recipes to Lower HIGH BLOOD PRESSURE

Delicious Vegetarian Recipes Based on Superfoods to Manage Hypertension

LA FONCEUR

Contents

Introduction	**4**
Cucumber Jaljeera	5
Summer Fruit Punch	6
Banana Choco Milk Shake	7
Crispi Garlici Sweet Potato	8
Fruit Chaat	10
Cucumber Honey Salad	11
Instant Beetroot Idli	12
Mixed Veg Raita	13
Multigrain Beetroot Paratha	15
White Radish Stuffed Palak Paratha	16
Rajma	18
Mushroom in Creamy Spinach Gravy	20
Veg Pulao	23
Dal Fry Palak Wale	25
Sweet Potato Burfi	26
Nutty Coffee Muffins	29
Red Velvet Halwa	30
Chocolate Shrikhand	31
Read More from La Fonceur	32
Connect with La Fonceur	32
About the Author	33

INTRODUCTION

Blood pressure is the measure of the force of blood against blood vessel walls. The blood pressure increases when the force of blood against the artery walls is too high, and this condition is known as hypertension or high blood pressure. To prevent and control hypertension, your diet should be rich in foods that have the following therapeutic effects:

· Foods that have diuretic effects.
· Foods that have potent vasodilating properties.
· Foods rich in magnesium as magnesium is a natural calcium channel blocker.
· Foods rich in potassium because potassium negates the sodium effect.
· Foods that keep nitric oxide levels high in your body.

Superfoods that prevent and control hypertension are:

Vasodilator: Beetroot and **garlic**.

Diuretics: Cucumber and **lemon**.

Magnesium-rich foods: Spinach, kidney beans, pumpkin seeds, pistachios, and **sweet potato**.

Potassium-rich foods: Cucumber, sweet potato, kidney beans, banana, spinach, and **curd**.

Foods that increase **nitric oxide production** in the body: **Beetroot, honey,** and **pumpkin seeds**.

This cookbook helps you include superfoods in your diet that help prevent and manage hypertension effectively.

CUCUMBER JALJEERA

INGREDIENTS

Lemon juice: 2 tbsp
Mint leaves: 50 g
Coriander leaves: 30 g
Ginger: 1½ inches
Green chili: 1
Cumin seeds: 1 tsp
Dried mango powder: ½ tsp
Asafoetida: ½ tsp
Black salt: To taste
Black pepper: 10
Cloves: 4
Cucumber pieces: 4 tbsp
Water: 1 L

DIRECTIONS

1. Dry roast cumin seeds, cloves, and black pepper till cumin seeds turn reddish-brown. Cool and grind them.
2. Finely chop the cucumber. Keep aside for later.
3. Take all the ingredients in the blender jar except the cucumber. Add about 50 ml - 100 ml of water and grind to make a super-fine paste.
4. Take the green paste in a jar and add more water to make 1 L. Add cucumber pieces. Refrigerate it for 2 hours. Serve chilled.

SUMMER FRUIT PUNCH

INGREDIENTS

Musk melon: 500 g
Cucumber: 250 g
Orange: 200 g
Mint leaves: 50 g
Honey: 1 tbsp/ To taste
Lemon juice: 1 tbsp
Dry mango powder: 1 tsp
Ice cubes: 8-10 (optional)

DIRECTIONS

1. Chill musk melon, cucumber, orange, and mint leaves in the refrigerator for 3-4 hours.
2. Keep 4-5 mint leaves to garnish. Blend musk melon, cucumber, orange and mint leaves, lemon juice, and honey in a blender.
3. Pour the fruit punch into glasses. Sprinkle dry mango powder over it. If you want, put 2-3 ice cubes in each glass. Enjoy the chilled and refreshing Summer Fruit Punch.

BANANA CHOCO MILK SHAKE

INGREDIENTS

Banana: 4
Chilled cow milk: 800 ml
Cocoa powder: 1 tbsp
Honey: 1 tbsp (optional)
Pistachios: 2 tbsp
Pumpkin seeds: 1 tbsp

DIRECTIONS

1. Dry roast pistachios and pumpkin seeds till they start releasing an aromatic smell and turn slightly brown. Remove from flame and let the nuts cool down. Chop them roughly using a knife or crush them using the mortar pestle or grinder.
2. Chop banana roughly. Blend banana, cocoa powder, and chilled milk into a smooth milkshake. Pour into glasses and add honey if needed.
3. Add pistachios and pumpkin seeds. Mix well and serve.

CRISPI GARLICI SWEET POTATO

INGREDIENTS

Sweet potato: 600 g
Rock salt: To taste
Kashmiri red chili powder: 1 tbsp
Black pepper powder: 1 tsp
Garlic: 15-18 cloves
Oil: 2 ½ tbsp
Corn flour: 1 tbsp
Dry mango powder: ½ tsp

DIRECTIONS

1. Wash sweet potatoes thoroughly. Peel them or leave the skin on. Cut them with a zig-zag knife or simple knife to medium thickness.

2. Steam sweet potatoes in a steamer for 5 minutes, not more than that.

3. Crush garlic finely using mortar pestle. Add Kashmiri red chili powder, salt, black pepper powder, and oil. Mix well.

4. Take out sweet potatoes in a colander. Add corn flour and toss to coat sweet potatoes evenly.

5. Add the garlic seasoning. Toss until sweet potatoes are evenly coated with the seasoning and don't look dry.

6. Divide sweet potatoes into 2 batches. Place one batch on a greased baking dish. Make sure they do not overlap each other.

7. If you have the grill function in your oven, grill sweet potatoes for 12 minutes. Flip the sides and grill again for 3-5 minutes.

8. Alternatively, bake the sweet potatoes in a pre-heated oven at 200 °C for 15 minutes. Flip the sides for even cooking and bake for 5-10 minutes or until sweet potatoes are crisp and browned from corners.

9. Take them out from the oven. Sprinkle dry mango powder and enjoy Crispi Garlici Sweet Potato with bell pepper chutney.

FRUIT CHAAT

INGREDIENTS

Thick curd: 250 g
Apple: 2
Banana: 2
Pomegranate: 1
Orange: 1
Papaya: 50 g
Kiwi: 2
Musk melon: 50 g
Almond: 10
Pistachios: 10
Walnut: 3 kernels
Pumpkin seeds: 3 tsp
Cumin: 1 tsp
Mint leaves: 2 tbsp
Honey: 1 tbsp
Salt: To taste
Dry mango powder: ¼ tsp

DIRECTIONS

1. Dry roast cumin seeds till they change color. Coarsely grind them.
2. Smoothen the curd by blending it with a spoon for 2 minutes.
3. Chop all the ingredients. Add them in curd. Mix well.
4. Add cumin seeds, dry mango powder, salt, and honey to it. Mix well. Refrigerate for 1 hour. Serve chilled.

CUCUMBER HONEY SALAD

INGREDIENTS

Cucumber: 1
Banana: 1
Sprouted brown chickpeas: 3 tbsp
Pomegranate: 3 tbsp
Peanuts: 2 tbsp
Red chili powder: ½ tsp
Honey: 1 tsp
Lemon juice: 2 tbsp
Chopped mint leaves: 2 tbsp
Black Salt: To taste

DIRECTIONS

1. Soak peanuts in water for 2 hours.
2. Cut banana and cucumber into half-inch cubes.
3. Roughly chop mint leaves and coriander leaves.
4. Take all ingredients in a bowl and mix well. Cover and leave for 15 minutes until the lemon juice moistens the salad.
5. Enjoy the Cucumber Honey Salad.

INSTANT BEETROOT IDLI

INGREDIENTS

Semolina: 200 g
Grated beetroot: 100 g
Curd: 125 g
Water: 50 ml to 100 ml
Salt: To taste
Fruit Salt: ½ tsp
Oil: For greasing

For coconut chutney
Freshly grated coconut: 200 gm
Coriander leaves: 100 gm
Roasted peanuts: 3 tbsp
Lemon juice: 2 tbsp
Ginger: ½ inch
Garlic: 4 cloves
Cumin: 1 tsp
Water: 50 ml

DIRECTIONS

1. Dry roast semolina for 5 minutes. Add curd and water to the semolina to make a thick paste. Leave aside for 15 to 20 mins.
2. Add grated beetroot and salt. Mix well. The batter should be thick, not runny, but if the batter is too thick, add two tablespoons of water.
3. In a pressure cooker/steamer, add 300 ml of water, cover with the lid, and let a boil come. Turn the flame to low. Grease the idli molds with oil.
4. Just before steaming, add fruit salt to the batter and mix well.
5. Pour the batter into the idli molds. Place the idli stand in the cooker/steamer. Close the lid and steam the idli for 12 to 15 minutes on medium flame.
6. Serve the instant beetroot idli with coconut chutney.

For Coconut Chutney
1. Grind together coconut, coriander leaves, peanuts, cumin, ginger, and garlic. Add water and blend again to make thick chutney.
2. Take it in a bowl. Add salt and lemon juice. Mix well

MIXED VEG RAITA

INGREDIENTS

Curd: 400 g
Grated beetroot: 2 tbsp
Onion: 50 g
Tomato: 50 g
Cabbage: 50 g
Cucumber: 50 g
Brown sugar: 1 tbsp
Black salt: 1 tsp
Cumin powder: 1 tbsp
Red chili powder: ½ tsp

DIRECTIONS

1. Finely chop onion, tomato, cucumber and cabbage.
2. Blend the curd until smooth. Mix all veggies except beetroot in it.
3. Add black salt, brown sugar, red chili powder, and cumin seeds powder to the curd. Mix well.
4. Keep the raita in the fridge for half an hour.
5. Garnish with grated beetroot. Enjoy Mixed Veg Raita with multigrain beetroot paratha.

MULTIGRAIN BEETROOT PARATHA

INGREDIENTS

Beetroot: 150 g
Bottle gourd: 150 g
Whole wheat flour: 150 g
Oat flour: 150 g
Gram flour: 75 g
Salt: To taste
Amaranth flour: 25 g
Coriander powder: 1 tsp
Ginger garlic paste: 1 tbsp
Red chili powder: 1 tsp
Jaggery: 1 tbsp
Turmeric: ½ tsp
Garam masala: 1½ tsp
Sesame seeds: 1½ tbsp
Asafoetida: a pinch
Olive oil: 1 tbsp+ as needed
Crushed fenugreek seeds: 1 tsp
Curd: 2 tbsp (to knead)

DIRECTIONS

1. Take all ingredients along with one tbsp of olive oil in a bowl.
2. Gradually add curd and knead to a stiff dough.
3. Take a dough ball, dip it in the dry whole wheat flour, and dust off the excess flour. Use a rolling pin to roll the dough into a circle.
4. Heat the pan/griddle/skillet (Tawa) on medium-high flame.
5. Place the paratha on the skillet. Cook for about a minute or cook until the paratha begins puffing up from the base at some places.
6. Flip the paratha and spread 3-4 drops of olive oil. Cook for 2 minutes until it turns light brown.
7. Flip the paratha again and top with 3-4 drops of olive oil, spread it evenly over the surface. Press the paratha gently with a flat spatula.
8. Once you begin to see brown spots on both sides of the paratha, transfer the paratha to a serving plate. Your paratha is ready. Similarly, make all the parathas.
9. Enjoy Multigrain Beetroot Paratha with mixed veg raita.

WHITE RADISH STUFFED PALAK PARATHA

INGREDIENTS

For Stuffing

White radish: 500 gm
Chopped green garlic: 50 g
Chopped ginger: 1 tsp
Chopped coriander leaves: 50 g

Cumin seeds: 1 tsp
Asafetida: ½ tsp
Salt: To taste

For Paratha

Whole wheat flour: 400 g
Lemon juice: 1 tbsp
Olive oil: 1 tbsp + for making paratha

Spinach: 200 g
Salt: To taste

DIRECTIONS

For Stuffing

1. Wash white radish thoroughly. Peel and grate radish. Sprinkle salt over it. Cover and keep aside for 15 minutes.
2. Squeeze radish to remove water. Take out as much water as you can. Keep the water for kneading.
3. Dry roast cumin seeds and crush them in a mortar with the help of a pestle.
4. Add cumin seeds, ginger, asafetida, green garlic, and coriander leaves to grated radish. Mix well.

For Paratha

1. Blanch the spinach with lemon juice and salt for 1 to 2 minutes using minimal water. Make a fine paste.
2. Add spinach paste and oil to whole wheat flour. Knead to a regular dough using radish water (if required). Cover and leave for 15 minutes.
3. Take a medium-sized ball from the dough. Make a smooth crack-free ball using your palms. Flatten the dough in your palm with your fingers. Keep center part a little thicker and sides thinner.
4. Put a spoonful of radish stuffing in the center, seal the edges, and make a smooth round ball. Flatten it gently with fingers. Dip it in dry whole wheat flour, dust off the excess flour. Use a rolling pin to roll it into a circle.
5. Heat the pan/griddle/skillet (Tawa) on medium-high flame. Place the paratha on the skillet. Cook for a minute.
6. Flip the paratha and spread 3-4 drops of olive oil. Cook for 2 minutes until it turns light brown.
7. Flip the paratha again and pour 3-4 drops of olive oil on top, and spread it evenly over the surface.
8. When brown spots appear on both sides of the paratha, take out the paratha on a serving plate. Your paratha is ready. Similarly, make all the parathas.

RAJMA

INGREDIENTS

Kidney beans: 300 g
Onion: 4 medium (1+3)
Tomato: 4 medium
Chopped ginger: 1 tbsp
Chopped garlic: 1 tbsp
Asafoetida: 1 tsp
Bay leaf: 1
Cumin seeds: ½ tsp
Garam masala: 1½ tbsp
Coriander powder: 1 tsp
Turmeric powder: ½ tsp
Red chili powder: 1 tsp
Salt: To taste
Mustard oil: 2 tbsp
Water: 900 ml
Coriander leaves: 20 g

DIRECTIONS

1. Soak kidney beans in enough water overnight.
2. Cut onion lengthwise. Keep one onion for cooking kidney beans and the rest three for gravy.
3. Pressure cook kidney beans with one chopped onion, salt, and water for 3-4 whistle. Check whether the kidney beans are cooked; If not, pressure cook for another 2 whistles.
4. Strain the cooked kidney beans and keep the stock for gravy.
5. Heat mustard oil in a pan. Once the oil is hot, add bay leaf, asafoetida, and cumin seeds.
6. When cumin seeds start changing color, add ginger and garlic and cook on medium flame for 2 minutes.
7. Add onions and cook for 6-8 minutes till onions turn translucent.
8. Add chopped tomatoes and salt. Mix well and cover and cook at low flame for 10 minutes. Stir occasionally.

9. Add garam masala, coriander powder, red chili powder, turmeric powder, and mix well. Cover and cook on low flame for 5 minutes.
10. Add kidney beans and cook for 5 mins on medium flame.
11. Add stock and bring it to a boil. Pressure cook for 1-2 whistles.
12. Garnish with coriander leaves and enjoy Rajma with rice and chapati.

MUSHROOM IN CREAMY SPINACH GRAVY

INGREDIENTS

Button mushrooms: 200 g
Melon seeds: 20 g
Coriander powder: ½ tsp
Garam masala powder: 1 tsp
Red chili powder: ½ tsp
Garlic: 10 cloves
Turmeric: ¼ tsp
Asafoetida: ¼ tsp
Lemon juice: 1 tsp
Ghee: 2 tbsp

Spinach: 200 g
Cashew: 30
Tomato: 3
Onion: 4
Ginger: 1 ½ inch
Bay leaf: 1
Cumin: ½ tsp
Salt: To taste
Water: 250 ml

For Tadka
Kashmiri red chili powder: ½ tsp
Chopped garlic: 1 tbsp
Ghee: 1 tsp

Asafoetida: ¼ tsp
Asafoetida: ¼ tsp

DIRECTIONS

1. Wash spinach thoroughly. Discard the stems and use fresh spinach leaves only. Blanch the spinach in 100 ml of water with salt and lemon juice for 1 minute. Let it cool and blend to a smooth paste.
2. Soak cashew nuts and melon seeds in 100 ml of hot water for 15 minutes. Grind to a fine white paste.
3. Grind together onion, ginger, and garlic to a fine paste. Separately blend tomatoes without adding water.
4. Cut mushroom into 1-inch cubes. Heat ½ tsp of ghee. Add mushroom pieces and sauté for 10 minutes.
5. Heat ghee in another pan. Add asafoetida, bay leaf, and cumin seeds. When cumin seeds start changing color, add

the onion-ginger-garlic paste. Cover and cook on low flame for 10 minutes till the raw taste of onion goes away completely.

6. Add tomato paste and salt. Cover and cook for 5 mins.

7. Add turmeric powder, garam masala, coriander powder, and red chili powder. Cover and cook on low flame for 10 minutes. Stir occasionally.

8. Add cashew-melon paste. Cover and cook on low flame for 15 minutes till the gravy leaves oil.

9. Add spinach paste and mix well. Bring it to a boil. If the gravy is too thick, add 50 to 100 ml water. Simmer for 5 minutes on low flame.

10. Add mushrooms. Mix and simmer for 2 minutes.

For Tadka

1. Add asafoetida, chopped garlic, and red chili powder to the hot ghee. Cook for 2-3 minutes.

2. Add tadka to mushroom in creamy spinach gravy. Cover and leave for 5 minutes. Serve with chapati and rice.

VEG PULAO

INGREDIENTS

Basmati rice: 400 g
Peas: 150 g
Cabbage: 150 g
Carrot: 150 g
Cloves: 6
Black pepper: 10
Cinnamon: 1 inch
Green cardamom: 3
Cumin seeds: 1 tsp
Asafoetida: ½ tsp
Bay leaf: 1
Salt: To taste
Ghee: 2 tbsp
Water: 850 ml

DIRECTIONS

1. Wash rice in running water and soak rice for 15 minutes.
2. Chop cabbage and carrot.
3. Crush clove, cardamom, cinnamon, and black pepper in a mortar.
4. Heat ghee in a pressure cooker. Add bay leaf, Asafoetida, cumin seeds, crushed cloves, cinnamon, cardamom, and black pepper. Cook till cumin seeds start changing color.
5. Add peas and salt, cover with a lid, and cook for 5 minutes.
6. Add carrots and cook for 5 minutes without covering. Add cabbage and cook for 5 minutes without covering.
7. Strain the rice and add it to the vegetable mixture. Cook for 6-8 minutes. Stir occasionally to prevent rice from sticking at the bottom of the cooker. Do not stir vigorously; otherwise, the rice will break.
8. Add water and close the lid of the cooker. Cook on medium flame for 1 whistle. Turn off the flame. Leave for 5 minutes and open the lid. Do not keep the lid on for a long time; otherwise, the rice will become soggy.
9. Enjoy Mixed Veg Pulav as it is or with rajma or dal fry.

DAL FRY PALAK WALE

INGREDIENTS

Spinach: 200 g
Yellow split pigeon peas: 140 g
Split Bengal gram: 50 g
Salt: To taste
Water: 650 ml
Turmeric: ½ tsp

For Tadka
Ginger: 1 inch
Garlic: 6 cloves
Onion: 1 medium
Tomatoes: 1 medium
Garam masala: ½ tsp
Coriander powder: ½ tsp
Red chili powder ¼ tsp
Cumin seeds: ½ tsp
Asafoetida: ½ tsp
Oil: 1 tbsp
Water: 250 ml

DIRECTIONS

1. Wash and chop the spinach leaves. Wash split pigeon peas and split Bengal gram 2-3 times and soak them in water for 15 minutes.

2. Discard the water. Take dal in the pressure cooker, add fresh 650 ml water. Bring to a boil. Add salt, turmeric powder, and spinach leaves. Close the lid and pressure cook on a medium flame for 4 whistles.

3. Heat oil in a pan. Add Asafoetida and cumin seeds. Cook till cumin seeds start changing color.

4. Add chopped ginger and garlic and cook for 2 minutes. Add onion and cook for 5-7 minutes. Add tomato and salt. Cook for 5 minutes.

5. Add garam masala, coriander powder, and red chili powder. Mix well. Cover and cook for 10 minutes. Mash the mix with a spatula.

6. Add dal to it and mix. Add 250 ml water, if required, add more water. Bring to a boil. Cook on low flame for 5-10 minutes. Serve hot with rice.

Note: Spinach absorbs less salt, due to which you will need less salt than usual, so add salt accordingly.

SWEET POTATO BURFI

INGREDIENTS

Sweet potato: 300 g (purple variety)
Jaggery: 60-70 g
Water: 100 ml
Ghee: 1 tbsp
Almond: 10-15

DIRECTIONS

1. Steam cook sweet potatoes. Remove skin and mash them using a potato masher or spoon.
2. Put jaggery in a pan. Heat it on low flame. When it starts melting, add water to it.
3. Turn the flame to medium and let the mix boil till *one string chashni* is formed. To check, put a drop of jaggery chashni on a plate. Let it cool down for 15 seconds. Take it between your thumb and index finger and check if it forms a string while separating your fingers. If it forms a string, it means it is done. If not, cook for another minute and check again.
4. Turn the flame to low and add mashed sweet potatoes and mix well. Keep stirring it continuously so that it does not stick to the bottom. Once mixed well, cook it on low to medium flame.
5. It will start forming a ball. At this stage, add 1 tbsp of ghee. Mix well till it stops sticking to the bottom and forms a non-sticky ball.
6. Turn off the flame and let it cool down a bit.

7. Take it out on the greased work platform and knead it for 5 minutes to make the smooth burfi.
8. Put butter paper on a flat plate and spread the dough on it. Roll it by pressing with palm or roll it with the help of a rolling pin. Keep the thickness from 0.5 cm to 0.75 cm. Let it cool down.
9. Dry roast 10-15 almonds. Open them in the vertical half.
10. Grease your palm with ghee and pat the top of the burfi. Cut in squares and stick half an almond on each piece. Keep in refrigerator and consume within 2 days.

NUTTY COFFEE MUFFINS

INGREDIENTS

Oats flour: 200 g
Amaranth flour: 75 g
Coffee: 2 tbsp
Milk: 100 ml
Sweet potato: 100 g
Banana: 1
Dates: 75 g
Raisins: 20 g
Water: 100 ml
Oil: 20 ml
Chopped pistachios: 15 pieces
Pumpkin seeds: 2 tbsp
Baking powder: 1 tsp
Baking soda: 1 tsp
Vanilla essence: ½ tsp

DIRECTIONS

1. Take oats flour and amaranth flour in a bowl. Pass the flour mixture three times through a sieve to make soft and smooth muffins. Keep aside.
2. Mix coffee powder with two tbsp of hot milk. Keep aside. Add a spoonful of oats flour to chopped pistachios and pumpkin seeds and coat well. This will prevent nuts from sinking into the batter.
3. Peel and cut sweet potatoes roughly. Pressure cook sweet potatoes, dates, and raisins in 100 ml of water for 3 whistles. Alternatively, you can boil them in a pan with the lid on. Once cool, add banana and blend to a smooth paste. Add milk while blending if required.
4. Take the prepared wet mix in a bowl. Add milk, coffee paste, vanilla essence, and oil. Mix well.
5. Gradually add the dry flour and mix well. Add chopped nuts and mix well.
6. Lastly, add baking powder and baking soda. Mix well and immediately pour the batter into paper-lined muffin cups and bake at 180°C for 25-30 minutes.

Tip: Adding baking powder and baking soda just before baking ensures that eggless muffins rise well and become spongy.

RED VELVET HALWA

INGREDIENTS

Grated Beetroot: 300 g
Desiccated coconut: 6 tbsp
Chopped mixed nuts: 2 tbsp

Milk: 500 ml
Jaggery: 50 g
Ghee ½ tsp

DIRECTIONS

1. Heat ghee in a pan. Add chopped nuts and roast till they start releasing a pleasant aromatic smell. Make sure to stir continuously. Take out the nuts and keep them aside.
2. In the same pan, add milk and bring it to a boil. Keep boiling till the milk thickens slightly.
3. Add grated beetroot to the milk. Keep the flame medium-high and let it cook till all milk evaporates. Keep stirring it. It will take about 15 to 20 mins.
4. When the milk has reduced to about 90%, it will look like a slurry. Add jaggery and desiccated coconut at this stage.
5. Cook for 5 minutes till the beetroot absorbs all the milk and starts leaving the pan. Turn off the flame.
6. Add all nuts. Serve hot or refrigerate for 3 hours and enjoy the Red Velvet Halwa.

CHOCOLATE SHRIKHAND

INGREDIENTS

Fresh thick curd: 2 L
Cocoa powder: 2 tbsp
Honey: To taste
Vanilla essence: ½ tsp
Dark chocolate pieces: 4 tbsp
Dark chocolate shaves: 2 tbsp

DIRECTIONS

1. Tie thick curd in a muslin cloth and hang it on a tap overnight or for 6-7 hours.
2. Take out the hung curd in a bowl. Blend it with a hand blender to make it smooth.
3. Mix cocoa powder, honey, and vanilla essence in the hung curd.
4. Take shrikhand in a muslin cloth and pass the shrikhand through muslin cloth by squeezing the muslin cloth. Alternatively, pass the shrikhand through a sieve to make it smooth.
5. Add the chocolate pieces and mix well. Refrigerate it for 3-4 hours.
6. Take out Shrikhand in serving bowls. Decorate with chocolate shavings. Keep it in the refrigerator for half an hour and enjoy Chocolate Shrikhand after lunch.

READ MORE FROM LA FONCEUR

Hindi Editions

CONNECT WITH LA FONCEUR

Instagram: **@la_fonceur** | **@eatsowhat**

Facebook: **LaFonceur** | **eatsowhat**

Twitter: **@la_fonceur**

Follow on Bookbub: **www.bookbub.com/authors/la-fonceur**

Sign up to the websites to get exclusive offers on La Fonceur eBooks:

Health Blog: **http://www.eatsowhat.com/signup**

Website: **www.lafonceur.com/sign-up**

ABOUT THE AUTHOR

With a Master's Degree in Pharmacy, the author La Fonceur is a Research Scientist and Registered Pharmacist. She specialized in Pharmaceutical Technology and worked as a research scientist in the pharmaceutical research and development department. She is a health blogger and a dance artist. Her previous books include *Eat to Prevent and Control Disease*, *Secret of Healthy Hair*, and *Eat So What!* series. Being a research scientist, she has worked closely with drugs and based on her experience, she believes that one can prevent most of the diseases with nutritious vegetarian foods and a healthy lifestyle.

Ingram Content Group UK Ltd.
Milton Keynes UK
UKHW020654190323
418778UK00016B/255/J